I0030445

Interactive
Mathematics Program®

INTEGRATED HIGH SCHOOL MATHEMATICS

As the Cube Turns

FIRST EDITION AUTHORS:
Dan Fendel, Diane Resek, Lynne Alper, and Sherry Fraser

CONTRIBUTORS TO THE SECOND EDITION:
Sherry Fraser, IMP for the 21st Century
Jean Klanica, IMP for the 21st Century
Brian Lawler, California State University San Marcos
Eric Robinson, Ithaca College, NY
Lew Romagnano, Metropolitan State College of Denver, CO
Rick Marks, Sonoma State University, CA
Dan Brutlag, Meaningful Mathematics
Alan Olds, Colorado Writing Project
Mike Bryant, Santa Maria High School, CA
Jeri P. Philbrick, Oxnard High School, CA
Lori Green, Lincoln High School, CA
Matt Bremer, Berkeley High School, CA
Margaret DeArmond, Kern High School District, CA

Key Curriculum Press

Second Edition   I M P

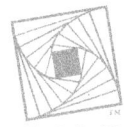

This material is based upon work supported by the National Science Foundation under award numbers ESI-9255262, ESI-0137805, and ESI-0627821. Any opinions, findings, and conclusions or recommendations expressed in this publication are those of the authors and do not necessarily reflect the views of the National Science Foundation.

Key Curriculum Press
1150 65th Street
Emeryville, California 94608
email: editorial@keypress.com
www.keypress.com
10 9 8 7 6 5 4 3 2 1   15 14 13 12 11
ISBN 978-1-60440-148-6
Printed in the United States
of America

Project Editors
Mali Apple, Josephine Noah

Project Administrator
Emily Reed

Professional Reviewers
Rick Marks, Sonoma State University, CA
D. Michael Bryant, Santa Maria High School, CA, retired

Accuracy Checker
Carrie Gongaware

First Edition Teacher Reviewers
Kathy Anderson, Aptos High School, CA
Dan H. Brutlag, Tamalpais High School, CA
Robert E. Callis, Hueneme High School, CA
Susan Schreibman Ford, Delhi High School, CA
Mary L. Hogan, Arlington High School, MA
Jane M. Kostik, Patrick Henry High School, MN
Brian Lawler, California State University San Marcos, CA
Brent McClain, Vernonia School District, OR
Michelle Novotny, Eaglecrest High School, CO
Barbara Schallau, East Side Union High School District, CA
James Short, Oxnard Union High School District, CA
Kathleen H. Spivack, Wilbur Cross High School, CT
Linda Steiner, Orange Glen High School, CA
Marsha Vihon, Corliss High School, IL
Edward F. Wolff, Arcadia University, PA

First Edition Multicultural Reviewers
Genevieve Lau, Ph.D., Skyline College, CA
Luís Ortiz-Franco, Ph.D., Chapman University, CA
Marilyn Strutchens, Ph.D., Auburn University, AL

Copyeditor
Brandy Vickers

Interior Designer
Marilyn Perry

Production Editor
Andrew Jones

Production Director
Christine Osborne

Editorial Production Supervisor
Kristin Ferraioli

Compositors
Kristin Ferraioli, Maya Melenchuk

Art Editor/Photo Researcher
Maya Melenchuk

Technical Artists
Lineworks, Inc., Maya Melenchuk, Kristin Ferraioli

Illustrator
Juan Alvarez, Alan Dubinsky, Tom Fowler, Nikki Middendorf, Briana Miller, Evangelia Philippidis, Paul Rodgers, Sara Swan, Martha Weston, April Goodman Willy, Amy Young

Cover Designer
Jenny Herce

Printer
Lightning Source, Inc.

Executive Editor
Josephine Noah

Publisher
Steven Rasmussen

CONTENTS

As the Cube Turns—Programming and Transformational Geometry

As the Cube Turns

Programming and Transformational Geometry

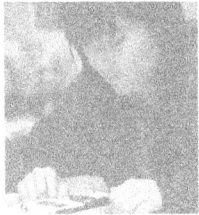

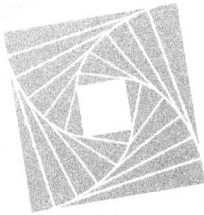

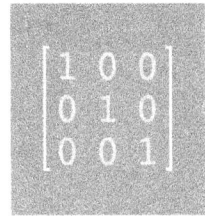

$$\begin{bmatrix} 1 & 0 & 0 \\ 0 & 1 & 0 \\ 0 & 0 & 1 \end{bmatrix}$$

As the Cube Turns—Programming and Transformational Geometry

Calculator Pictures

Have you ever wondered how animators create all of those amazing effects you see on the movie screen? In this unit, you will learn some of the mathematics behind computer animation. You will also create your own animation programs on your graphing calculator.

You will begin your investigations by discovering how to use your calculator to draw pictures.

Stephanie Wood and Elizabeth Graf begin the unit by conferring about how to draw pictures on the calculator.

Picture This!

Your goal today is to learn about drawing pictures on your graphing calculator, which is the first step toward creating an animation of a turning cube on the calculator.

Begin by experimenting with the calculator's keys and menus. You may want to read parts of the calculator manual or look up ideas in its index or table of contents. You might also ask other students for ideas.

Take careful notes on your own discoveries and on ideas you get from other students. Your notes might be helpful later, when you work on the turning-cube program or on your project.

"A Sticky Gum Problem" Revisited

Do you remember Ms. Hernandez and her twins? They were the main characters in a Year 1 POW called *A Sticky Gum Problem* (in the unit *The Game of Pig*). You will now examine a variation on that problem.

The Original POW

To refresh your memory, here is the original scenario.

> Every time Ms. Hernandez passes a gumball machine, her twins each want a gumball. They also insist on having gumballs of the same color. Gumballs cost a penny each, and Ms. Hernandez has no control over which color will come out of the machine next.

First, Ms. Hernandez and the twins passed a gumball machine with only two colors. Then they came across a machine with three colors. In each case, you found out the most money Ms. Hernandez might have to spend to satisfy her twins.

Then another parent, Mr. Hodges, came by the three-color gumball machine. He had triplets, so of course he needed three gumballs of the same color. You found the most money he might have to spend to satisfy his triplets.

Finally, you generalized the problem by finding a formula that would work for any number of colors and any number of children. Your formula would tell you the maximum amount a parent might have to spend to provide each of his or her children with a gumball of the same color.

1. Before starting on the new problem, re-create the generalization for the old one.

   a. Find a formula for the maximum amount a parent might have to spend in terms of the number of colors and the number of children.

   b. Provide a proof of your generalization. That is, give a convincing argument to show that your formula is correct.

continued

Some New Gumball Problems

Ms. Hernandez's twins have grown up a bit in the last three years, and they have changed in some ways. Now each insists on getting a gumball of a color *different* from the other twin's. Gumballs are still a penny each.

One day, Ms. Hernandez and the twins pass a machine that contains exactly 20 gumballs: 8 yellow, 7 red, and 5 black. As before, Ms. Hernandez cannot control which gumball will come out of the machine.

2. If each twin wants one gumball, what's the maximum amount Ms. Hernandez might have to spend so that the twins get different colors? Prove your answer.

3. Because the twins have grown, they sometimes each want two gumballs. The two gumballs each twin gets must be the same color, so the flavors match, but one twin's gumballs have to be a different color from the pair the other twin gets.

   What's the maximum amount Ms. Hernandez might have to spend? Prove your answer.

New Generalizations

Now make at least three generalizations about these new problems. You get to decide exactly what you want to generalize. Here are some options:

- Generalize the number of children.
- Generalize the number of gumballs each child wants.
- Generalize the number of gumballs of each color.
- Generalize the number of colors of gumballs in the machine.

continued

Your formula or procedure should tell how to find the maximum amount a parent might have to spend to provide all the children with the particular number of gumballs, with each child getting gumballs of a different color.

Write a proof for each generalization you create. As a grand finale, try to generalize all of these variables.

Write-up

Your write-up for this problem should begin with your formula and proof for Question 1 and your answers (with proofs) for Questions 2 and 3.

Then present each generalization you found for the new type of problem, with a proof for each generalization. Also explain how you arrived at each generalization and how you discovered your proof.

Adapted from "A Sticky Gum Problem" in *aha! Insight* by Martin Gardner (New York: W. H. Freeman and Company, 1978).

Starting Sticky Gum

Read the Problem of the Week *"A Sticky Gum Problem" Revisited.*

Assume that a parent comes across a machine containing many gumballs of several colors and wants to give each of several children a gumball of the same color.

1. This is Question 1 of the POW.

   a. Find a formula for the maximum amount a parent might have to spend in terms of the number of colors in the gumball machine and the number of children.

   b. Provide a proof of your generalization. That is, give a convincing argument to show that your formula is correct.

Suppose Ms. Hernandez and the twins pass a gumball machine that contains 20 gumballs: 8 yellow, 7 red, and 5 black.

2. This is Question 2 of the POW. If each twin wants one gumball, what's the maximum amount Ms. Hernandez might have to spend so that the twins each get a different color? Prove your answer.

3. Identify any questions you have about what is expected in the POW.

Programming Without a Calculator

Every programming language uses very specific syntax and commands. This formal programming language is often called *programming code*. The code for a particular task is likely to vary from one calculator model to another.

Programmers often begin with "plain-language descriptions" of what they want to do and then turn those descriptions into programming code. Several activities in this unit will ask you to write or interpret plain-language programs. Each of these programs will begin with a title line, and each command will begin on a new line.

1. Read through the steps of the plain-language program. Then draw on graph paper what should appear on the calculator screen when the program is run. Assume the calculator has an appropriate viewing window.

    Program: LINES

    Clear the screen

    Draw a line segment connecting $(-4, -2)$ to $(2, -2)$

    Draw a line segment connecting $(-4, 2)$ to $(-1, 3)$

    Draw a line segment connecting $(-1, 3)$ to $(1, 5)$

    Draw a line segment connecting $(2, -2)$ to $(4, 0)$

    Draw a line segment connecting $(1, 5)$ to $(4, 4)$

    Draw a line segment connecting $(1, 5)$ to $(-2, 4)$

    Draw a line segment connecting $(-4, 2)$ to $(-2, 4)$

    Draw a line segment connecting $(4, 4)$ to $(4, 0)$

    Draw a line segment connecting $(-1, 3)$ to $(2, 2)$

    Draw a line segment connecting $(-4, 2)$ to $(-4, -2)$

    Draw a line segment connecting $(2, 2)$ to $(4, 4)$

    Draw a line segment connecting $(2, -2)$ to $(2, 2)$

    If you don't get something that looks like a real picture, you've probably made a mistake somewhere and should check your work.

continued ▶

2. Turn the plain-language program into programming code. In other words, write the program lines you would enter into your calculator to produce your drawing from Question 1.

3. Add more commands to your program to improve the rather dull picture. Then draw what should appear on the screen when you run your improved program.

Programming Loops

You may think of a loop as something you make with a shoelace or a piece of string, but it's also a handy device for writing computer and calculator programs. **Loops** allow you to repeat the same set of instructions in a program as many times as you like.

Over the next several activities, you will use loops to simplify the work of writing a program and to create the illusion of motion.

Rebecca Yaeger writes a report about her work so far in determining how to rotate a cube.

Learning the Loops

The For/End combination of instructions can be used to do lots of interesting things. Your job is to figure out how it all works, both in the plain-language version and in the version using programming code for your calculator.

To show the overall structure of the program more clearly, it's helpful to indent the body of the loop. This is done in the plain-language programs in this and later activities. In addition, each step of the body of the loop starts with a bullet (the symbol •).

1. Describe what will happen when you run a calculator program based on this plain-language program.

Program: LOOP1

For T from 1 to 5

   • Display "HELLO" on the screen

End the T loop

continued ▶

2. Describe what will happen when you run a calculator program based on this plain-language program.

Program: FRUITLOOP

For A from 1 to 5

- Display "LEMON" on the screen

- For B from 1 to 3

  - Display "LIME" on the screen

- End the B loop

End the A loop

3. Sketch what the calculator should show when you run a program based on this plain-language description.

Program: LOOP2

For G from 3 to 10

- Draw a line segment from (G, G) to $(G + 3, G + 1)$

End the G loop

4. Write a short program (in programming code, using a For/End loop) that you think will draw something interesting. Describe in words what you think your program will draw.

An Animated Shape

You will now write a program to create and animate a shape on your calculator. As you progress, write down any questions you have about writing programs for animation. Keep in mind that you are trying to create the illusion of motion.

1. First, draw your shape on graph paper. Make your shape very simple, with no more than five line segments or pieces.

2. Next, write out the program code to have the calculator draw your shape. You may want to begin with a plain-language program.

3. a. Draw the same shape on graph paper in a new location, near the first.

   b. Write additional program instructions to make it appear that your shape has moved to its new location.

   c. Repeat parts a and b several times, changing the position of your shape slightly each time.

4. Enter your program into the calculator and run it. Locate and correct errors if the program doesn't run as you would like.

A Flip Book

A flip book is a device consisting of a series of pictures, each slightly different from the previous one. The pictures are drawn on index cards (or something similar), and you flip through them quickly to create the illusion of movement.

1. Create a simple flip book of your own. You don't have to be a great artist to do this; a moving rectangle or a rolling ball is fine. On the other hand, if you like drawing, have fun!

2. Explain how you think this activity is related to the unit.

Movin' On

It's often helpful to have a *setup program* at the beginning of a calculator program. The setup program might adjust the viewing window, clear the screen, and so on.

From now on, we will include a line that simply says "Setup program" in all plain-language programs that involve screen graphics. When you translate such programs into code, you will need to give details for the setup program.

1. a. Use drawings on graph paper to describe the result of this plain-language program.

   Program: SEGMENTS

   Setup program

   For S from 1 to 5

   • Clear the screen

   • Let T be 3 more than S

   • Let U be 6 more than S

   • Draw a line segment from (S, S) to (T, T)

   • Draw a line segment from (T, T) to (U, S)

   • Draw a line segment from (U, S) to (S, S)

   • Delay

   End the S loop

   b. Create programming code for the plain-language program. Include an appropriate setup program. Write the setup program using plain-language instructions, if necessary.

2. Make a general list of the components you would include in a setup program. This might include items that were not needed in Question 1.

Some Back and Forth

1. This sequence of graphs shows a line segment going back and forth between two positions. Use a loop to write a plain-language program to produce an animation showing a line segment in these positions, one position after the other. Your program does not need to draw the coordinate axes or scales. They are shown here merely to indicate the changing position of the line segment.

First position of segment

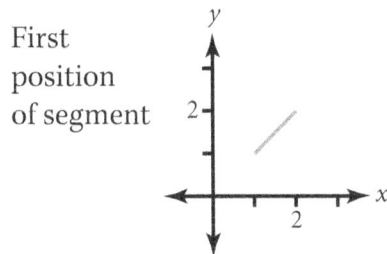

Second position of segment

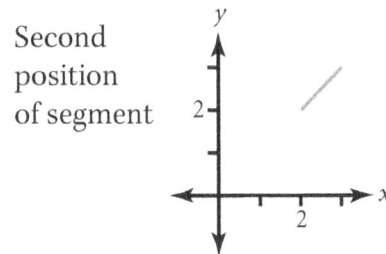

Third position of segment

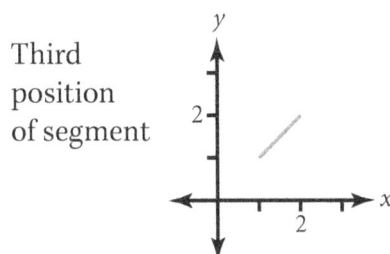

Fourth position of segment

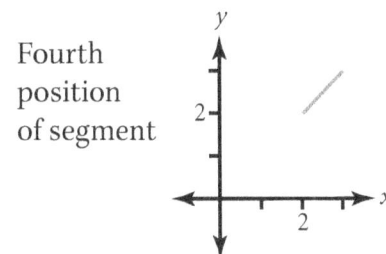

Fifth position of segment

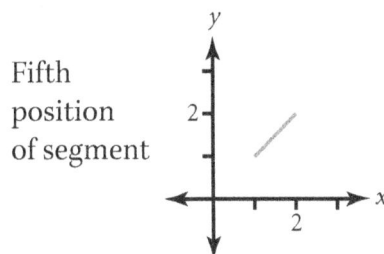

Sixth position of segment

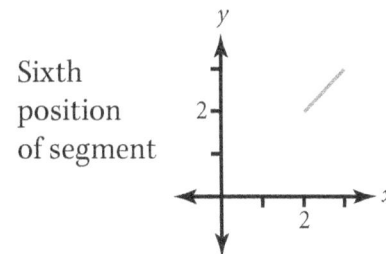

Seventh position of segment

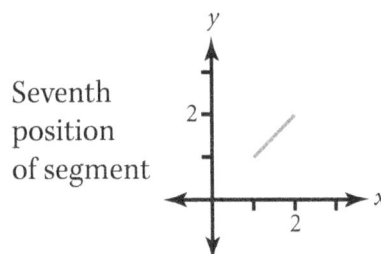

Eighth position of segment

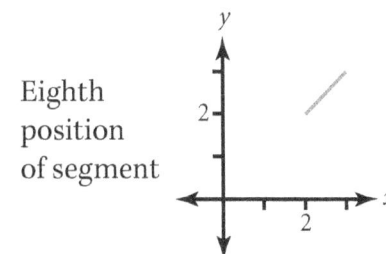

2. Create programming code for your plain-language program, including an appropriate setup program.

Arrow

This plain-language program describes the animation of a flying arrow. Create programming code for this plain-language program. Enter the code into your calculator, run the program, and then find ways to improve it.

Program: ARROW

Setup program

For A from 1 to 8

- Clear the screen

- Make B 3 less than A

- Make C 1 less than A

- Draw a line from (B, B) to (A, A)

- Draw a line from (C, A) to (A, A)

- Draw a line from (A, C) to (A, A)

- Delay

End the A loop

Sunrise

You will now create a program that displays a rising sun by showing a circle "moving" up and across the screen. There are three parts to this activity.

- Make a sequence of drawings on graph paper to plan what you want to appear on the screen. Show at least four different positions for the circle.

- Write a plain-language program describing how to create your sequence of drawings.

- Write programming code for your plain-language program. Include settings for the viewing window so that your circles will be visible.

A Wider Windshield Wiper, Please

Imagine! Straight out of high school, you land a job at Better Design Ideas, Inc. Your supervisor is just starting to explain the company's project to design a windshield wiper that will clean more area than the standard wiper.

The standard wiper consists of a 12-inch blade rigidly attached to a 12-inch arm. The middle of the blade is at the end of the arm, and the arm rotates back and forth, making a 45° angle with the horizontal at each end of its motion. The shaded area in the first diagram shows the path swept out by the blade of the standard wiper.

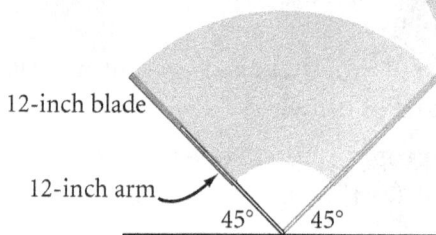

12-inch blade

12-inch arm

45°   45°

While your supervisor is talking, two men in white coats rush in with a new model. "It's better!" exclaims one. "It's worse!" shouts the other.

The new model also has a 12-inch blade attached at its midpoint to a 12-inch arm. But in this case, the blade pivots so that it's always vertical. The arm rotates 90° as in the standard model. The next diagram shows the path swept out by the blade of the new wiper.

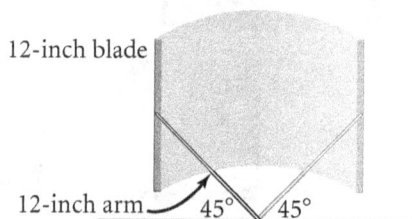

12-inch blade

12-inch arm   45°   45°

The two men keep shouting, but your supervisor merely rolls her eyes, hands you a white coat, and tells you to find out who's right. Just as you are racing out the door, she calls after you, "In any case, make a better one—one that will clean more area."

You know you are limited to 12-inch blades, which are an industry standard, and that any rotating arm can rotate only 90°.

As your write-up for this POW, write a report for your supervisor. Be sure to state any further assumptions you make.

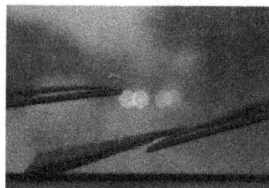

Translation in Two Dimensions

You have now learned the principles of how to create the appearance of motion as well as how to create programming code to carry out some basic animation. Part of this work involved translating plain-language programs into programming code.

Now you will work with another meaning for the word *translation*—a geometric meaning—that will help you carry out a specific animation task. You'll also find out how to use matrices to assist with this task.

Rachel Silverman has successfully programmed her calculator to make a straight line appear animated.

Move That Line!

Your task in this activity is to write a program that will make a line segment move across the screen by repeating the same **translation.** After you have done this for a single segment, you will do it for a more complex picture.

1. a. Draw a line segment on graph paper.

   b. Choose a translation. Draw the result of applying that translation to your line segment.

   c. Apply the same translation to the segment you got in part b. Apply it again to your new result, and so on, until you have applied the translation five times altogether.

2. a. Write a plain-language program that will produce the results you drew on graph paper in Question 1.

   b. Write programming code for your plain-language program.

   c. Enter and test your programming code, modifying it if necessary.

3. If you didn't use a loop in Question 2, redo your programs using a loop. Use variables for the coordinates of the endpoints, and use steps before you start the loop to set the initial values of these variables. Then think about how the coordinates for each new pair of endpoints are found from the coordinates of the preceding pair, and put steps in your loop to make these changes in the variables.

4. Once you know how to write a program using a loop to make a single line segment move, make a simple picture out of four or five segments. Then write a program to make that whole picture move through several identical translations.

Double Dotting

In Questions 1 and 2, read the plain-language program and show on graph paper what should appear on the calculator screen.

1. **Program: DOTS**

    Setup program

    Set A equal to 3

    Set B equal to 2

    Set C equal to 9

    Set D equal to 2

    For P from 1 to 7

    - • Draw dots at (A, B) and (C, D)

    - • Increase A, B, and D by 1 each

    - • Decrease C by 1

    End the P loop

2. **Program: MOREDOTS**

    Setup program

    Set A equal to 3

    Set B equal to 5

    Set C equal to 3

    Draw a dot at (A, B)

    For P from 1 to 5

    - • Decrease A and B by 1

    - • Increase C by 1

    - • Draw dots at (A, B) and (C, B)

    End the P loop

3. Write a program of your own that uses variables in loops to draw dots.

Memories of Matrices

It turns out that matrices can be helpful in expressing geometric translations. Do you remember matrices? In case you need a review, this activity begins with part of an activity from the Year 3 unit *Meadows or Malls?*

1. A matrix could be used to keep track of students' points in a class. Each row could stand for a different student. The first column might be for homework, the second for oral reports, and the third for POWs.

   Suppose this table represents results for the first grading period:

	Homework	Reports	POWs
Ana	18	54	30
Ben	35	23	52
Cass	46	15	60
Devon	60	60	60

   A matrix representation of this information might look like this:

   $$\begin{bmatrix} 18 & 54 & 30 \\ 35 & 23 & 52 \\ 46 & 15 & 60 \\ 60 & 60 & 60 \end{bmatrix}$$

   Here, in table form, are the students' points in each category for the second grading period:

	Homework	Reports	POWs
Ana	10	60	0
Ben	52	35	58
Cass	42	20	48
Devon	60	60	60

continued ▶

a. Write these second-grading-period scores in a matrix.

b. Compute each student's total points *in each assignment category* for the two grading periods combined. Write the totals in matrix form.

c. In part b, you added two matrices. Based on your work, write an equation showing two matrices being added to give your matrix from part b.

Now take a look at how matrix addition relates to translations.

2. Suppose you have a diagram of a house set up in a coordinate system like the one shown here.

a. Make a matrix A out of the coordinates of the five points that are at the corners of the house, using a separate row of the matrix for each point.

Now suppose you want to translate this diagram, moving it so that the lower left corner of the house ends up at (4, 2).

b. Make a matrix B out of the coordinates of the translated house.

c. Find a matrix C so that [A] + [C] = [B]. Matrix C is called a *translation matrix*.

3. Let's use the letter A for the matrix shown in Question 1. Matrix A has 4 rows and 3 columns, so it has 12 entries altogether.

$$A = \begin{bmatrix} 18 & 54 & 30 \\ 35 & 23 & 52 \\ 46 & 15 & 60 \\ 61 & 60 & 60 \end{bmatrix}$$

You will need to be familiar with notation used to represent individual entries of a matrix. Mathematicians use the symbol a_{ij} to refer to the entry in the *i*th row and *j*th column (with the lowercase letter *a* indicating that the entry is from the matrix represented by uppercase A). For example, a_{32} refers to the entry in the third row and second column of matrix A, so a_{32} is 15.

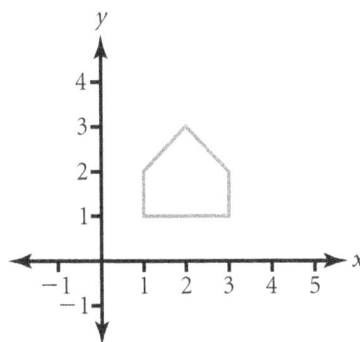

continued ⬇

The context makes clear that the numbers 3 and 2 here are separate subscripts, not the number 32. For larger matrices, you would need to use commas to separate the subscripts. For instance, $a_{12,34}$ would represent the entry in the 12th row and 34th column of a larger matrix A.

a. What is a_{23} in the matrix A shown above?

b. Write the entry 30 from matrix A in the form a_{ij}.

c. Write a new matrix B with entries defined by this set of equations:

$$b_{11} = 3 \qquad b_{12} = 6 \qquad b_{13} = 2$$
$$b_{21} = 5 \qquad b_{22} = 1 \qquad b_{23} = 3$$

Cornering the Cabbage

Did you ever read about Peter Rabbit and his siblings, Flopsy, Mopsy, and Cotton-tail?

Actually, it doesn't matter whether you did or not. All you need to know is that Peter used to steal cabbages from a local farmer, Mr. McGregor. But now Peter and Flopsy are working for Mr. McGregor.

Peter and Flopsy usually get paid in cabbages, but Mr. McGregor decides that because they've been working so hard, he'll pay them by giving them some land. This will allow them to grow some of their own cabbages. Peter and Flopsy think this is a great idea.

Two poles are lying nearby, and Mr. McGregor gives one to Peter and one to Flopsy. He tells them to hold them together at the ends and spread them out at any angle they want. Then he will connect the two other ends, forming a triangle. They will get all the land inside the triangle.

It turns out Peter's pole is approximately 2 meters long and Flopsy's is approximately 3 meters long. Of course, they want the most land they can get. They aren't sure what angle to choose or even if it matters. Can you help them?

1. a. Figure out what the area would be if the angle formed by the two poles were 25°.

   b. Figure out what the area would be if the angle were 100°.

2. Pick two other angles to try. Figure out what area Peter and Flopsy will get if they use each of those angles.

3. Develop a general formula for the area of the triangle. Use t and u for the lengths of the two given sides and θ for the angle they form, as in the diagram.

4. Determine what value for the angle θ will give Peter and Flopsy the largest area, and justify your answer.

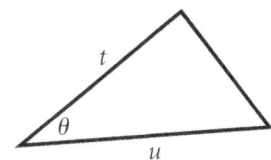

Rotation in Two Dimensions

The next step in turning a cube in three dimensions is learning to turn things in two dimensions. To accomplish this, you will need to develop some trigonometric formulas. You probably won't be surprised to learn that with rotations, as with translations, matrices can be very useful.

Jeff Tung, Ken Hoffman, and Audrey Rae use coordinates to help them work in two and three dimensions.

Goin' Round the Origin

Being able to rotate pictures around the origin will be useful when you want to create interesting graphics. And, of course, **rotation** will be important in turning a cube.

You will begin that task by examining a specific problem and its generalization.

1. Consider point P in the diagram, with coordinates $(3, 4)$. Suppose this point is rotated $10°$ counterclockwise around the origin to the position shown as point Q.

   What are the coordinates of point Q? You might find trigonometry useful here.

2. Now consider the point R, with coordinates $(2, 5)$. Rotate this point $10°$ counterclockwise around the origin. What are the coordinates of its new location?

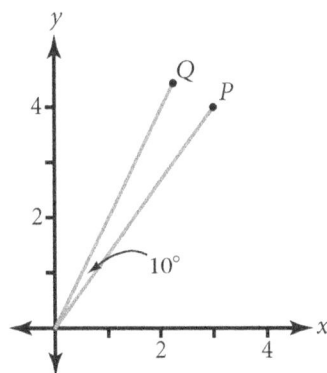

3. Generalize Questions 1 and 2. Begin with a point with coordinates (x, y). Rotate that point $10°$ counterclockwise around the origin. Develop a formula or procedure for getting the coordinates of the new point in terms of x and y.

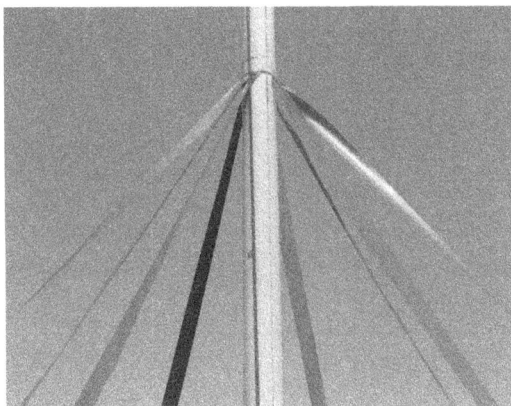

Double Trouble

Do you remember Woody from the *Shadows* unit in Year 1? As you may recall, Woody was very fond of measuring trees. He was especially happy to learn how to use trigonometry so he could avoid climbing trees.

One day, Woody leaned a ladder between the branches so that its tip just reached the top of a tree. Next, he measured the angle that the ladder made with the ground. He wrote down the angle and headed home to do the computation to find the tree's height. (He knew the length of the ladder.)

When he got home, he grabbed his scientific calculator, punched in sin 40°, and got 0.643. Then he punched in cos 40°, just in case he might want to know how far the foot of the ladder had been from the tree. He found that cos 40° was about 0.766.

When he did the calculation, his answer seemed way off. He ran back to the scene and measured the angle again. Lo and behold, it was really 80°!

When Woody returned home, he couldn't find his calculator anywhere. His friend Elmer suggested that Woody simply multiply sin 40° by 2 to get sin 80°.

1. Is Elmer's suggestion correct? That is, is sin 80° equal to 2 sin 40°?

2. How could you decide if Woody was right if you had no way to look up sin 80°?

In *Cornering the Cabbage,* you saw that the area of a triangle such as the one shown here is given by the expression $\frac{1}{2} tu \sin \theta$. Use this formula in Question 3.

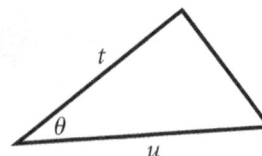

continued ▶

3. The first diagram below shows an isosceles triangle with two sides of length x and an angle of 80° formed by those sides. The second diagram shows the same triangle with the altitude drawn. The altitude has length z and splits the original triangle into two smaller triangles.

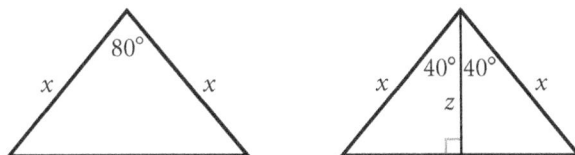

a. Find the area of the original triangle (in terms of x).

b. Find the area of each of the two smaller triangles (in terms of x and z).

4. a. Use the fact that the large triangle is made up of the two smaller triangles to get an expression for sin 80°. This expression may involve sin 40° and both x and z.

b. Work with your answer to part a to get an expression for sin 80° that involves neither x nor z, but may involve both sin 40° and cos 40°.

The Sine of a Sum

You have noted that when a point is rotated $10°$ around the origin, the rectangular coordinates for its new location involve the expressions $\cos(\theta + 10°)$ and $\sin(\theta + 10°)$.

It would be helpful if you knew how to find the sine or cosine of two angles added together, using their individual sines and cosines. In this activity, you'll begin with sine.

Your task is to develop a formula for $\sin(A + B)$, where A and B are two angles. You can use $\sin A$, $\cos A$, $\sin B$, and $\cos B$ in your formula. You might consider modifying your work on *Double Trouble* using a diagram like the one shown here.

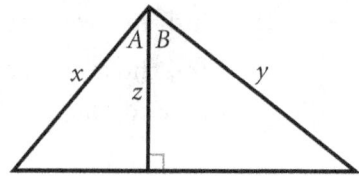

A Broken Button

Elmer's phone rang. He knew it was Woody before he even picked it up. Ever since Woody had lost his calculator, he'd been pestering Elmer.

This time, Woody wanted to know the value of cos 50°. But when Elmer went to find out on his own calculator, he discovered that the cosine key was broken.

Fortunately, the sine key was working. Elmer offered to give Woody the value of sin 50° if that would help. Woody wrote down the value of sin 50°, hung up, and proceeded to find cos 50° using the **Pythagorean identity,** $\sin^2 x + \cos^2 x = 1$.

1. Show how Woody might have used the Pythagorean identity and the numeric value of sin 50° to determine the value of cos 50°.

Later, Elmer called to tell Woody that he'd found an easy, no-computation way to get cos 50°, by using a right triangle and simply finding the sine of a different angle.

2. How do you think Elmer's method works?

Oh, Say What You Can See

You may have recently used the trigonometric identity $\cos \theta = \sin (90° - \theta)$. Recall that an *identity* is an equation that holds true no matter what values are substituted for the variables.

Your task now is to give examples illustrating this and other identities and then to explain each identity in one or both of these ways:

- Using the situation of the Ferris wheel (from *The Diver Returns*).
- Using the graphs of the equations $y = \sin \theta$ and $y = \cos \theta$, which are shown here for your reference.

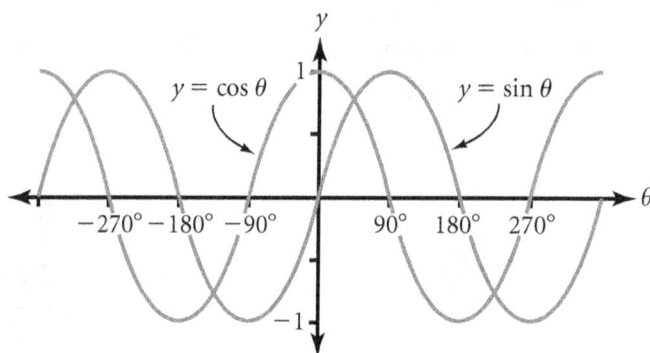

1. Consider the equation $\cos \theta = \sin (90° - \theta)$.
   a. Choose two values for θ, in different quadrants, and verify that the equation holds true for those values.
   b. Write an explanation of the equation in terms of the Ferris wheel model.
   c. Write an explanation of the equation in terms of the graphs of the sine and cosine functions.

2. Consider the equation $\cos (-\theta) = \cos \theta$.
   a. Choose two values for θ, in different quadrants, and verify that the equation holds true for those values.
   b. Write an explanation of the equation using either the Ferris wheel model or the graph of the cosine function.

continued ◗

3. Consider the equation $\sin(-\theta) = -\sin\theta$.

a. Choose two values for θ, in different quadrants, and verify that the equation holds true for those values.

b. Write an explanation of the equation using either the Ferris wheel model or the graph of the sine function

Comin' Round Again (and Again . . .)

To write a calculator program to rotate pictures, you need to be able to rotate points.

You have seen that when a point (x, y) is rotated counterclockwise about the origin through an angle ϕ, the coordinates (x', y') of its new location are given by the equations

$$x' = x \cos \phi - y \sin \phi$$

$$y' = x \sin \phi + y \cos \phi$$

To understand how to use these formulas, first draw a small picture on a coordinate grid. Use only line segments in your picture to connect points. Use at least five points to get an interesting picture.

1. Write the x- and y-coordinates of the points in your picture.

2. Rotate your picture 60° around the origin by rotating each point. Don't guess about the new coordinates. Calculate them and show your work. Round off to the nearest tenth.

3. Based on your answers to Question 2, draw the rotated picture. Be sure to keep track of which pairs of points you should connect with line segments.

4. Rotate your new picture another 60° as you did previously, showing your work.

More Memories of Matrices

In *Memories of Matrices,* you reviewed the addition of matrices and saw how to apply that idea to the geometric translation of points.

Now it's time to review some more about matrices and then use them for rotations. Once again, the first question is based on an activity from *Meadows or Malls?*

1. Lindsay uses a transport plane to deliver goods for two customers, Charley's Chicken Feed and Careful Calculators. Here are some important facts about her customers:

   - Charley's Chicken Feed packages its product in containers that weigh 40 pounds and are 2 cubic feet in volume.

   - Careful Calculators packages its materials in cartons that weigh 50 pounds and are 3 cubic feet in volume.

   a. Organize this information into a matrix, using the first row for Charley's Chicken Feed and the second row for Careful Calculators. Call this matrix A.

   b. On Monday, Lindsay transports 500 containers of chicken feed and 200 cartons of calculators. Put those facts into a row matrix (that is, a matrix with one row). Call this matrix B.

   c. Use the information in your two matrices to find out the total weight the plane carried and the total volume used. Put those two answers into a row matrix. Call this matrix C.

continued ▶

Congratulations! If you completed Question 1c, you have multiplied two matrices. The matrix C is the product [B] · [A].

2. Based on your work, write a description of how to multiply matrices. Include a statement about when it is possible to multiply two matrices.

3. In Question 1, you found the matrix product [B] · [A]. Now try to find the matrix product [A] · [B]. Did you get the same result as in Question 1? Explain.

4. On Tuesday, Lindsay transports 400 containers of chicken feed and 300 cartons of calculators. On Wednesday, she transports 250 containers of chicken feed and 350 cartons of calculators.

   a. Make a matrix showing Lindsay's transportation record for all three days, using a different row for each day. Call this matrix D.

   b. Describe how you would use matrix multiplication to get a matrix showing the weight and volume the plane carried each day.

5. Use the idea of matrix multiplication to express rotations. That is, find a way to use matrices to get from the ordered pair (x, y) to the ordered pair $(x \cos \phi - y \sin \phi, x \sin \phi + y \cos \phi)$.

Taking Steps

You have seen how For/End instructions can be used to create a programming loop in which the loop variable increases by 1 each time the program goes through the loop.

For example, consider this plain-language program:

> For A from 4 to 8
>
> > • Show A on the screen
>
> End the A loop

This program will print the numbers 4, 5, 6, 7, and 8 on the calculator screen.

It's possible to make the loop variable increase by something other than 1. For example, you can probably use an instruction on your calculator that looks something like this:

> For C from 2 to 11, step 3

This instruction will start the variable C at 2 and then increase it by 3 each time the program comes back to this instruction. The number 3 is called the *increment* or *step value*.

1. What do you think will appear on the screen when this program is run?

   Program: STEP3

   > For C from 2 to 11, step 3
   >
   > > • Show C on the screen
   >
   > End the C loop

2. Write a plain-language program that will show the calculator counting by 5s from 30 to 50.

continued ▶

3. What do you think would happen in each of these programs?

a. **Program: MISSTEP**

   For W from 2 to 10, step 3

   • Show W on the screen

   End the W loop

b. **Program: FRACSTEP**

   For S from 3 to 6, step 0.7

   • Show S on the screen

   End the S loop

4. How might you get a calculator to count backward? For example, write a plain-language program (using a For/End loop) that you think would display the values 5, 4, 3, 2, and 1, in that order.

How Did We Get Here?

In this unit, you have worked with several different mathematical concepts. Before you go on, reflect on where you have been and think about where you are going.

1. Make a list of the important mathematical terms and formulas you've learned or used so far in the unit. Write a definition of each of the key terms.

2. Write out the purpose of this unit as clearly as you can.

3. For each item you listed in Question 1, discuss how it relates to the purpose of this unit.

Swing That Line!

In this activity, your task is to write a calculator program that will take a line segment and rotate it counterclockwise around the origin a certain number of degrees. Your program should repeat this rotation several times and allow you to see each segment briefly before erasing it and showing the next one.

1. Make a careful sketch on graph paper of the lines you want the calculator to draw.

2. Write a plain-language program to create the animation. Use what you can of your *Move That Line!* program, simply making the necessary changes.

3. Turn your program into programming code for your calculator.

4. Enter and run your program from Question 3.

5. Once your program is successful, record both the plain-language program and the programming code on paper.

6. Now modify your program to work with a more complicated picture.

Doubles and Differences

You have developed formulas for the sine and cosine of the sum of two angles, writing sin $(A + B)$ and cos $(A + B)$ in terms of sin A, cos A, sin B, and cos B. Now you will develop some variations on those formulas.

1. Develop a formula for the sine of the difference between two angles. That is, find a formula for sin $(A - B)$ in terms of sin A, cos A, sin B, and cos B. It may be helpful to think of sin $(A - B)$ as sin $[A + (-B)]$.

2. Check your work by choosing several pairs of values for A and B and testing whether your formula works.

3. Find a formula for cos $(A - B)$ in terms of sin A, cos A, sin B, and cos B. Then check your formula by substituting pairs of values for A and B.

4. In *Double Trouble*, you used ideas about area to find a formula for sin 80° in terms of sin 40° and cos 40°. Now use the formula for sin $(A + B)$ to develop a formula for sin $(2A)$ in terms of sin A and cos A. Check your formula by substituting values for A.

5. Use the formula for cos $(A + B)$ to develop a formula for cos $(2A)$. Check your formula by substituting values for A.

What's Going On Here?

Once again, your task is to figure out what a certain plain-language program does. Show on graph paper what would appear on the calculator screen after someone executes the program MYSTERY.

Program: MYSTERY

Setup program

Clear the screen

Let A be the matrix $\begin{bmatrix} 2 & 2 \\ 5 & 2 \\ 2 & 6 \\ 5 & 6 \\ 3.5 & 9 \end{bmatrix}$

Let B be the matrix $\begin{bmatrix} \cos 30° & \sin 30° \\ -\sin 30° & \cos 30° \end{bmatrix}$

For C from 1 to 12

- Draw a line from (a_{11}, a_{12}) to (a_{21}, a_{22})

- Draw a line from (a_{21}, a_{22}) to (a_{41}, a_{42})

- Draw a line from (a_{41}, a_{42}) to (a_{51}, a_{52})

- Draw a line from (a_{11}, a_{12}) to (a_{31}, a_{32})

- Draw a line from (a_{31}, a_{32}) to (a_{51}, a_{52})

- Replace matrix A by the product $A \cdot B$

End the C loop

Projecting Pictures

You are now ready to move on to three dimensions. You will need to review some things you learned in *Meadows or Malls?* about three-dimensional graphs. You will also need to adapt some of the two-dimensional geometry you learned in the Year 3 unit *Orchard Hideout* to three dimensions.

Cassie Burniston and Sam Ellison explore the challenge of creating a two-dimensional drawing of a three-dimensional cube.

An Animated POW

In this POW, you will work with a partner to create an interesting animation program for the graphing calculator.

You need to hand in a general outline of the program, a more detailed plain-language program, and the actual code for the program. Be sure to keep a written copy of your work at all times in case your programs are accidentally deleted.

You and your partner will make a presentation of your animation to the class, lasting three to four minutes. You will show your animation and describe one interesting feature of how you did the programming.

Here are the stages of your work on this POW:

- Give your teacher a statement of who your partner is.

- With your partner, hand in a description of what you want your program to do.

- With your partner, make a presentation to the class that lasts three to four minutes. Your presentation should include

  ○ a demonstration of your program running on the overhead calculator

  ○ a description of one interesting feature of your program

- Hand in the written outline, plain-language program, and code.

"A Snack in the Middle" Revisited

You may recall from the Year 3 unit *Orchard Hideout* that Madie and Clyde sometimes spend their afternoons pruning trees. Each afternoon, they choose the two trees that need it the most.

Pruning makes them hungry, so they set up a snack table between the two trees they are working on. They set up the table at the midpoint of the line segment connecting the two trees.

1. If Madie and Clyde are working on the trees at (24, 6) and (30, 14), where should they set up the table? Explain your answer.

2. If they are working on the trees at (6, 2) and (11, 9), where should they set up the table? Explain your answer.

One afternoon, Clyde is daydreaming about a delicious snack, and he slips from his ladder and twists his knee. Madie graciously offers to do the pruning herself, but Clyde won't hear of it.

Madie also offers to move the snack table to whatever tree Clyde is working on, but he won't agree to that, either. Clyde finally agrees to place the table one-third of the way from himself to Madie, instead of halfway.

3. The next time they go out, Clyde is working on the tree at (6, 9) and Madie is working on the tree at (18, 15). According to their new agreement, where should they put the table? Explain your answer.

Fractional Snacks

In *"A Snack in the Middle" Revisited*, Clyde injured his knee. Because he was having trouble getting around, he and Madie decided to place their snack table closer to Clyde than to Madie.

This activity continues with the lives of our busy pruners on the following day.

1. In the morning, Clyde wakes up feeling even worse. So he and Madie decide to put the table only one-fourth of the way from him to Madie. If Clyde works on the tree at $(-2, 7)$ and Madie works on the tree at $(14, 19)$, where should they put the table?

2. Clyde begins to feel a little better. So the friends increase the fraction: Now the table will be two-fifths of the way from Clyde to Madie. This time, Clyde's position is $(6, 13)$ and Madie's is $(9, 1)$. Where should they put the table?

3. As Clyde's condition varies, he and Madie keep changing the fraction they use for placing the snack table. Describe how to compute the coordinates of the snack if the given "fraction of the distance" from Clyde to Madie is r.

More Walking for Clyde

Clyde's knee is pretty well healed, but he needs to walk regularly to strengthen his leg muscles. So Clyde and Madie have decided to place the snack table so that Clyde actually has to walk past Madie to get to it.

Assume Clyde is at (10, 12) and Madie is at (16, 2). The table will still be on the straight line that passes through their two positions.

1. Suppose Clyde and Madie decide to make the distance from Madie to the table equal to half of her distance from Clyde. Where should they put the table?

2. Suppose Clyde and Madie decide to make the distance from Madie to the table equal to twice her distance from Clyde. Where should they put the table?

3. In general, if the distance from Madie to the table is t times the distance from Madie to Clyde, where should they put the table?

Monorail Delivery

When Clyde recovers from his accident, his next project is to help Madie build a monorail in the orchard. A monorail is a train that runs along a single track.

Madie and Clyde plan to use the monorail to help them transport the apples they grow to a central location for shipping to market.

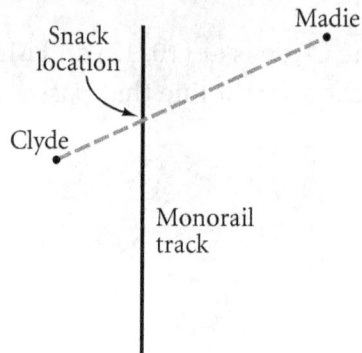

The first monorail they build runs along the line $x = 12$, and the train runs along this track in both directions. Clyde and Madie realize they could also use the monorail to bring themselves fresh snacks when they are out in the orchard pruning trees. The monorail would drop off their snacks at the point where the line segment that joins their locations crosses the track. (They would work on opposite sides of the track, and they have stopped worrying about who will be closer to the snack.)

1. If Clyde is working on the tree at (3, 7) and Madie is working on the tree at (30, 19), where should the monorail drop off the snack? Make a sketch of the situation and explain your answer.

2. If Clyde is at (−5, 28) and Madie is at (18, −14), where should the monorail drop off the snack? Make a sketch of the situation and explain your answer.

3. Generalize your work from Questions 1 and 2. If Clyde is at (a, b) and Madie is at (c, d), where should the monorail drop off the snack?

Another Mystery

In this activity, your job is to figure out what this plain-language program does and then create programming code for it. (Remember that the order of operations for matrix arithmetic is the same as the order of operations for number arithmetic.)

Program: ANOTHER

Setup program

Clear the screen

Let A be the matrix $\begin{bmatrix} 1 & 1 \\ 4 & 1 \\ 2 & 4 \end{bmatrix}$

Let B be the matrix $\begin{bmatrix} 6 & -1 \\ 6 & -1 \\ 6 & -1 \end{bmatrix}$

Let C be the matrix $\begin{bmatrix} \cos 180° & \sin 180° \\ -\sin 180° & \cos 180° \end{bmatrix}$

For W from 1 to 4

• Draw a line from (a_{11}, a_{12}) to (a_{21}, a_{22})

• Draw a line from (a_{11}, a_{12}) to (a_{31}, a_{32})

• Draw a line from (a_{21}, a_{22}) to (a_{31}, a_{32})

• Replace A by the matrix $B + A \cdot C$

End the W loop

---

1. Make a careful drawing on graph paper of what the screen will show after someone executes the program ANOTHER.

2. Create programming code for your calculator that will accomplish what this plain-language program describes.

A Return to the Third Dimension

In the Year 3 unit *Meadows or Malls?* you worked with the three-dimensional coordinate system to understand and solve linear programming problems. Let's review some ideas about that coordinate system.

1. Imagine a cube in the three-dimensional coordinate system, placed in a position that you choose. You may want to use a small cube, along with a cardboard version of the coordinate system, to create a model of the situation.

   a. Write down the coordinates of the eight vertices of your cube.

   b. A cube has six faces. Give the equations of the six planes that contain the faces of your cube.

   c. Sketch your cube in the three-dimensional coordinate system.

2. Write the equations of several pairs of planes that are parallel to each other.

3. Write the equations of a pair of planes that are neither parallel nor perpendicular to each other. Sketch your two planes.

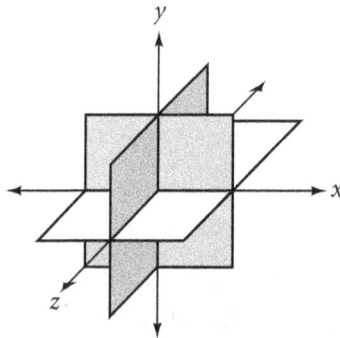

Where's Madie?

One day as he is pruning away, Clyde thinks of something he wants to ask Madie. He decides to take a break and walk over to see her.

He doesn't remember where she is working, but he knows where the snack is on the line segment between them. He also remembers Madie saying she would have to walk twice as far for her snack that day as he would.

1. If Clyde is at $(2, 6)$ and the snack is at $(12, 1)$, where is Madie?

2. Suppose you don't know the exact coordinates of either Clyde's tree or the snack, and assume the monorail still runs along the line $x = 12$. Create general instructions for finding the coordinates of the tree where Madie is working. Verify that your instructions work for the situation in Question 1.

3. Let Clyde's coordinates be (a, b) and the snack's coordinates be $(12, c)$. Write a formula for the coordinates of Madie's tree.

And Fred Brings the Lunch

Two spiders, Bonita and Charlotte, like to spin webs at opposite corners of a room. Bonita likes heights, so she always goes to one of the corners of the ceiling.

Charlotte does not care for heights, so she always goes for a corner of the floor. For proper artistic balance, she goes to the corner diagonally opposite from Bonita.

A single thread connects the spiders' webs. As the spiders work on their webs, they get hungry. So they have their associate, Fred the fly, bring in lunch each day. The spiders want their lunch partway between them, along the connecting thread.

Fred does not glide up or down threads as spiders do. He simply flies. So Bonita and Charlotte map out a coordinate system to tell him where to fly. They designate the rear left corner of the floor, where Charlotte usually hangs out, as (0, 0, 0).

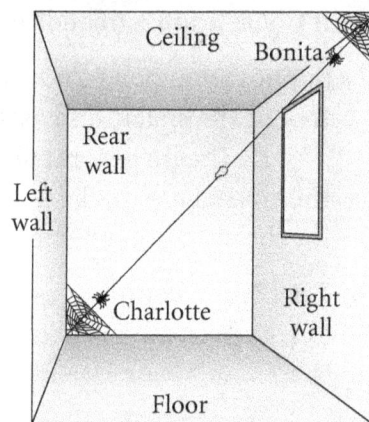

Bonita is in the front right corner of the ceiling. The spiders set up the system so that the coordinates for her position, (a, b, c), work this way:

- The first coordinate, a, represents how many feet to the right Bonita is from Charlotte.
- The second coordinate, b, represents how many feet up Bonita is from Charlotte.
- The third coordinate, c, represents how many feet forward Bonita is from Charlotte.

1. One day, the spiders are in a room that is 11 feet wide, 13 feet tall, and 24 feet long. If Charlotte is at (0, 0, 0) and Bonita is at the opposite corner at (11, 13, 24), and they want lunch halfway between them, where should Fred bring lunch?

continued

2. After she is used to this room, Charlotte gets adventurous and builds a web slightly out from her corner, at (2, 1, 4). Bonita also leaves her corner, building her web at (9, 12, 20). The connecting thread is still in a straight line between their webs, and the spiders still want lunch halfway between them. Where should Fred make his delivery?

3. Charlotte suddenly decides that with her distaste for heights, she should not have to go halfway up. She and Bonita compromise, agreeing that Charlotte will go one-third of the way up the thread from her position while Bonita will come two-thirds of the way down. Based on Bonita's and Charlotte's positions in Question 2, where should Fred go?

4. Using the positions from Question 2 again, where should Fred go if the lunch is to be placed two-fifths of the way from Charlotte to Bonita?

5. Charlotte and Bonita are tired of all this mental activity, figuring out what to tell Fred. They would like an equation to use instead.

   Assume Charlotte is at (x_1, y_1, z_1), Bonita is at (x_2, y_2, z_2), and r is a fraction between 0 and 1, so that they want lunch "r of the way" along the thread from Charlotte to Bonita. Create an equation in terms of these variables that states where Fred should bring lunch.

Flipping Points

You have learned about translations and rotations, and you've developed formulas to find the new coordinates after one of these **geometric transformations** is applied to a point.

Translations and rotations are two important examples of a special kind of transformation called an **isometry.** An isometry is a way of moving all the points in the plane (or in 3-space) so that the size and shape of objects are unchanged. The word *isometry* means "same measure," which means the distance between any two points doesn't change when the points are moved.

There is a third basic category of isometry called a **reflection.** Reflections are also known as *flips*. A reflection in the plane is defined by giving a *line of reflection*. The reflection moves each point P to a point Q so that the line of reflection becomes the perpendicular bisector of the segment connecting P and Q. In other words, Q is chosen so that $\overline{PQ}$ is perpendicular to the line of reflection and the line of reflection intersects $\overline{PQ}$ at its midpoint. Point Q is called the reflection of P across the line of reflection.

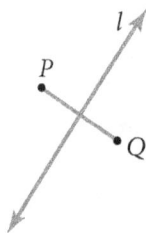

For example, in the diagram, l is the line of reflection, point Q is the reflection of point P across l, and l is the perpendicular bisector of $\overline{PQ}$. Thus, the reflection of a point is the mirror image of the original point across the line of reflection.

continued

The next diagram shows a triangle in the first quadrant and its reflection using the *y*-axis as the line of reflection.

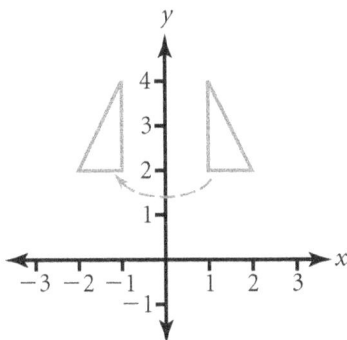

For instance, the reflection of the point $(1, 2)$ is the point $(-1, 2)$, and the line of reflection (the *y*-axis) is a line of symmetry between the original triangle and its reflection.

1. Using the diagram, give the coordinates for each of the other two vertices of the original triangle (in the first quadrant) and give the vertices of each of their reflections (in the second quadrant).

2. Generalize your results. That is, if (a, b) is an arbitrary point, what are the coordinates of its reflection across the *y*-axis?

3. Find a way to represent this transformation using a matrix process.

Where's Bonita?

Fred generally receives his instructions from Charlotte, so he knows what her coordinates are. Of course, he also knows where he is putting the lunch. But after a while, he starts wondering where Bonita is.

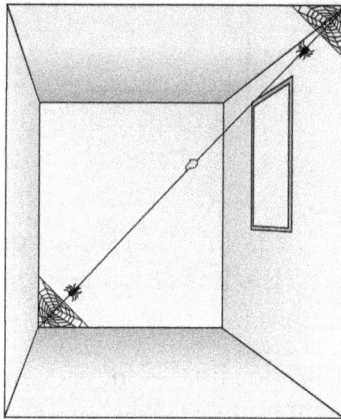

1. For example, one day, when Charlotte is at (1, 3, 2), she tells Fred that the lunch should be at (5, 8, 9) and that this is halfway to Bonita. Where is Bonita?

2. On another occasion, when Charlotte will go only one-third of the way toward Bonita to get lunch, Charlotte tells Fred to put the lunch at (4, 4, 7). Charlotte herself is at (2, 1, 1). Where is Bonita?

3. Fred's brain is really taxed one day when Charlotte tells him to put the lunch at (3, 6, 7) and explains that this will be two-fifths of the way to Bonita. Charlotte is at (1, 2, 2). Where is Bonita?

Lunch in the Window

The two spiders are still spinning away, but Bonita has developed claustrophobia and prefers to work outside. So the spiders find a room where one of the windows is kept open. Charlotte works inside, while Bonita does her spinning outside.

The thread connecting them goes through the open window, and Charlotte and Bonita will have their lunch right where the thread passes through the window opening. They need to figure out the coordinates of the point where the thread passes through the open window.

Charlotte and Bonita are using the same coordinate system as before, with the origin at the rear left corner of the floor. The first coordinate gives the distance to the right, the second gives the distance up, and the third gives the distance toward the front. The window is in the wall on the right side of the room, and the room is 14 feet wide from left to right.

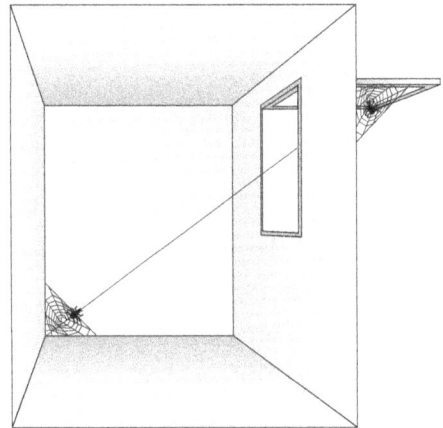

By the way, the spiders' associate Fred mysteriously disappeared after the last lunch. His son, Fred Jr., is now assisting them.

1. Before you start on the lunch problems, find the equation of the plane that contains the window.

2. Suppose Charlotte is at $(4, 2, 2)$ and Bonita is at $(24, 20, 8)$. Where should Fred Jr. bring their lunch? Explain your reasoning carefully.

3. Now suppose Charlotte is at $(10, 0, 4)$ and Bonita is at $(26, 16, 8)$. Where should Fred Jr. drop off lunch? Again, explain your reasoning.

4. This time Charlotte is at $(7, 0, 4)$, while Bonita is still at $(26, 16, 8)$. Give Fred Jr. directions about where to put lunch.

5. Generalize your results, based on the coordinates (x_1, y_1, z_1) for Charlotte and (x_2, y_2, z_2) for Bonita.

Further Flips

In *Flipping Points,* you looked at the isometry of reflecting figures across the *y*-axis, as shown in the first diagram here. But any line can be used as a line of reflection. In this activity, you will consider some other cases.

1. Start with the same original triangle, but this time use the line $y = x$ as the line of reflection, as shown in the second diagram. Find the coordinates of the vertices of both the original triangle and the reflected triangle.

2. Generalize Question 1 by finding the image of an arbitrary point (a, b) reflected across the line $y = x$.

3. Express this reflection in terms of matrices. That is, find a matrix process that will turn the vector $[a\ b]$ into the corresponding image vector when reflected across the line $y = x$.

4. Now repeat Questions 1 to 3 using the same original triangle and the line $x = 6$ as the line of reflection.

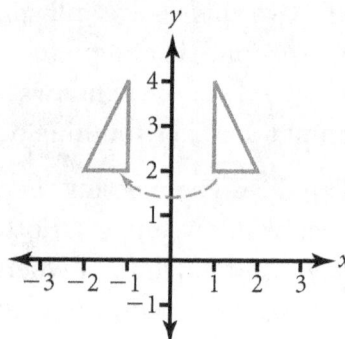

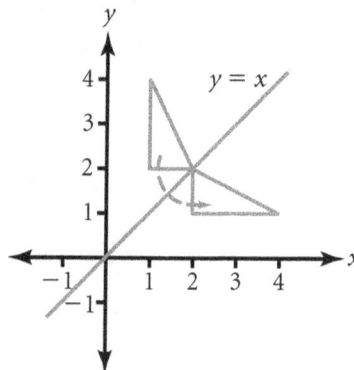

Cube on a Screen

As part of the unit problem, you need to figure out how to draw a cube, which is three-dimensional, on a calculator screen, which is two-dimensional. Such a drawing is called a **projection.**

This activity will give you some insight into how this can be done. You will work with a partner on this activity. You and your partner will need a cube, a sheet of clear plastic, and three pens of different colors. Think of the plastic as representing the calculator screen. The screen should be held vertically, like a plane parallel to the *xy*-plane.

In Part I, partner A will hold the cube and screen while partner B makes three sketches of the cube, as described in Questions 1, 2, and 3. In Question 4, you and your partner will compare the three sketches.

You'll switch roles in Part II. Partner A will make two sketches, as described in Questions 5 and 6. In Question 7, you and your partner will again compare the sketches.

Part I: Changing the Viewpoint, Moving the Cube

1. To begin with, partner A should take the cube and the screen and place them in some fixed position, holding the screen vertically between the cube and partner B. Partner B need not be directly in front of the cube.

   Without moving her head, partner B should look at one corner of the cube and place a dot on the screen where her line of vision crosses the screen. (Partner B may want to imagine a laser beam from her eye to the corner of the cube and place the dot where the beam would burn a hole in the screen.)

   Partner B should continue like this, without moving, imagining her line of sight tracing all the edges of the cube, and marking on the screen where her line of vision would cross the screen as the cube is traced. The result should be a two-dimensional drawing—a projection—of the three-dimensional cube.

continued ▶

2. When the drawing from Question 1 is complete, partner B should move so that she is in a different position compared to the cube. *The cube and screen should be left in the same position as in Question 1.*

   Partner B should now trace the cube from this new position, using the same "laser" method as in Question 1, but with a pen of a different color.

3. Next, partner A should move the cube closer to the screen. *Partner B should stay in the same position as in Question 2.* Partner B should make a third sketch of the cube, again using the laser method, but with the third pen.

4. Compare the drawings from Questions 1 to 3. How are they different? Has the drawing merely moved, or has it actually changed?

Part II: Rotating the Cube

Now switch roles, with partner A doing the drawing and partner B holding the cube. Hold the screen so that a clean portion of it is between partner A and the cube.

5. Partner A should make an initial sketch of the cube, as described in Question 1.

6. Partner A and the screen should stay in the same position while partner B moves the cube through a partial rotation (such as $45°$ or $90°$) around the z-axis. That is, imagine a line perpendicular to the screen, and have the cube do a partial rotation around this axis, turning as it goes around the axis as in the central unit problem.

   Partner A should sketch the cube in its new position, using a pen of a different color from that used in Question 5.

7. Compare the drawings from Questions 5 and 6. How are they the same, and how are they different? Has the drawing merely turned, or has it actually changed?

Spiders and Cubes

You've recently completed a series of activities involving the eating habits of the spiders Bonita and Charlotte.

- *And Fred Brings the Lunch*
- *Where's Bonita?*
- *Lunch in the Window*

This unit, though, is about programming a calculator to draw a turning cube. What do lunch arrangements for spiders have to do with the unit problem? Your task in this activity is to figure out and explain this connection.

Find Those Corners!

Let's fix a 2-by-2-by-2 cube in a three-dimensional coordinate system. For convenience, we'll place it snugly in the corner where the coordinate planes meet. Three of the cube's faces will be against the coordinate planes, with one vertex of the cube at $(0, 0, 0)$ and the diagonally opposite vertex at $(2, 2, 2)$.

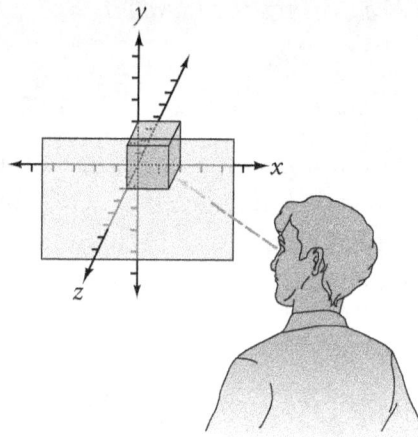

1. Find the coordinates of the other six corners.

2. Imagine that the plane $z = 5$ is a screen. Your teacher will give you a viewpoint. Write down the coordinates of this viewpoint.

   Using the plane $z = 5$ as the screen and the viewpoint you are given, determine the coordinates where each vertex of the cube will be projected onto the screen.

3. Plot the projections of the vertices you found in Question 2 on a large piece of graph paper, thinking of the piece of graph paper as the plane $z = 5$. Put an appropriate scale on the axes.

   Label the projections for the vertices $(2, 2, 2)$, $(0, 0, 0)$, and so on, and connect them to draw the cube. Use solid lines for the edges that are visible and dotted lines for the edges that are hidden by the rest of the cube.

   If your drawing doesn't look something like a cube, you will need to revise your work.

An Animated Outline

You and your partner will now complete the outline for your animation project, *An Animated POW.* You will turn it in tomorrow, but keep a copy so you can continue your work.

The next stage after this activity will be to write the program itself, beginning with a plain-language program. Remember to keep written copies of all your work on the program. It may be easier to find errors in your written work than on the calculator.

Plan to have your program entered well before your presentation day. Debugging your program will likely take more time than writing the first draft.

Mirrors in Space

You've looked at coordinate and matrix representations in two dimensions for each of the three basic types of isometries: translations, rotations, and reflections.

Now you will explore reflections in 3-space. In this setting, we use a *plane of reflection,* which is analogous to a line of reflection in two dimensions. The reflection of a point P across a plane m is the point Q that makes m the perpendicular bisector of $\overline{PQ}$. In other words, $\overline{PQ}$ is perpendicular to m, and m intersects the segment at its midpoint. The plane m will be a plane of symmetry between any set of points and the set of their reflections across m.

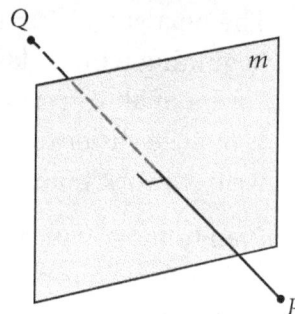

1. a. Begin with the point $(4, 6, 2)$, and imagine reflecting that point using the yz-plane as the plane of reflection. (The yz-plane is the same as the plane $x = 0$.) Determine where the point $(4, 6, 2)$ should end up.

   b. Generalize your work from part a, using the point (x, y, z). That is, find the coordinates of the point you would get if you reflected the point (x, y, z) across the yz-plane.

2. Find a matrix A that will do the work of Question 1b for you. That is, find a matrix A so that the matrix product $[x\ y\ z]\ [A]$ gives the coordinates of the reflection of the point (x, y, z) across the yz-plane.

Where Are We Now?

This is a good time to reflect on the unit. In this activity, you will describe the unit so far, and indicate what you think remains to be done.

1. Describe what your expectations were early in the unit. What did you think the unit would be about? What mathematics did you think you would be learning?

2. What have turned out to be the key mathematical ideas of the unit? How do they relate to the unit problem?

3. What ideas or procedures do you think you still need to learn in order to solve the unit problem?

Rotation in Three Dimensions

You are finally ready for the cube! You just need to combine what you know about rotations in two dimensions with what you know in general about three dimensions to create matrices for rotating in three dimensions. Then you will be turning the cube!

Joel Picazo uses coordinates and matrices to rotate objects in three dimensions.

Follow That Point!

1. Consider the point $(2, 3, -5)$. If this point is rotated $20°$ counterclockwise around the z-axis, where does it end up?

2. Now suppose you have a segment connecting $(2, 3, -5)$ and $(5, 2, 6)$ and you rotate that segment $20°$ counterclockwise around the z-axis. What is the result of this rotation?

One Turn of a Cube

Imagine placing a cube in 3-space. For simplicity, choose a placement so that the faces of the cube are parallel to the coordinate planes.

1. Write down the coordinates in 3-space for the eight vertices of your cube.

2. Imagine that the cube is rotated 30° counterclockwise around the z-axis. Choose one face of the cube, and find the new coordinates of the vertices on that face. Round off the coordinates to the nearest tenth.

3. Make two sketches, one showing your cube in its original position in 3-space and another showing its position after rotation.

Rotation Matrix in Three Dimensions

You have noted that if a point (x, y, z) is rotated counterclockwise around the z-axis through an angle of $30°$, its new coordinates are

$$(x \cos 30° - y \sin 30°, \ x \sin 30° + y \cos 30°, \ z)$$

1. Find a rotation matrix for this transformation. That is, find a matrix [B] so that

$$[x \ \ y \ \ z] \cdot [B] = [x \cos 30° - y \sin 30° \ \ x \sin 30° + y \cos 30° \ \ z]$$

2. What should the rotation matrix be if the rotation is around the x-axis? Around the y-axis?

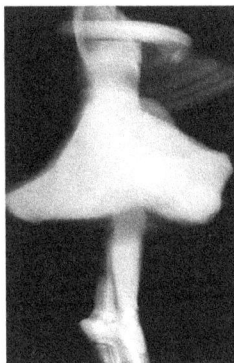

The Turning Cube Outline

Writing animation programs is a very complicated task, and the unit problem is no exception. You have had to learn many ideas about mathematics and programming to develop a program for the turning cube, and you now have all the necessary pieces.

Most programmers find it helpful to write an outline for a program before developing the code. Look over your work for this unit, and develop an outline for a program to turn the cube. You don't need to give line-by-line details. Instead, give the general structure of how the program will be organized.

Beginning Portfolio Selection

Now that you have turned the cube, you can understand the items on the outline you saw at the beginning of the unit. That outline probably looked something like this:

1. Draw a picture on the calculator.

2. Create the appearance of motion.

3. Change the position of an object located in a two-dimensional coordinate system.

4. Create a two-dimensional drawing of a three-dimensional object.

5. Change the position of an object located in a three-dimensional coordinate system.

Choose three items from this outline. For each item, select one activity that was important in developing that item. Explain how that activity helped you understand the given item and how that item fits into the overall development of the unit.

You will work with the remaining two items in *Continued Portfolio Selection.*

Creating Animations

Now that you have turned the cube, you are ready to create an interesting animation of your own. Over the next several days, you will put together your work on *An Animated POW*, present it, and see the animations your classmates have created.

Adé Thomas-Stewart and Joaquin Menjivar-Austerlitz work together to create an animation to present to their class.

An Animated POW Write-up

You and your partner have almost finished your work on *An Animated POW*. To complete the POW, you need to do three things:

1. Finish the program.

2. Prepare your written copy of the program.

3. Develop a presentation for the class that lasts three to four minutes. Your presentation should include

   • a demonstration of the program

   • a description of one interesting feature of the program

Continued Portfolio Selection

You will now continue the work on your portfolio that you began in *Beginning Portfolio Selection.* In that activity, you looked at the stages in the outline from the beginning of the unit. Here is the outline again.

1. Draw a picture on the calculator.

2. Create the appearance of motion.

3. Change the position of an object located in a two-dimensional coordinate system.

4. Create a two-dimensional drawing of a three-dimensional object.

5. Change the position of an object located in a three-dimensional coordinate system.

In the earlier activity, you selected three of the items from this outline. Now you will work with the remaining two items.

As before, for each of these items, choose one activity that was important in developing that item. Explain how the activity helped you understand the item and how the item fits into the overall development of the unit.

As the Cube Turns Portfolio

You will now put together your portfolio for *As the Cube Turns*. This process has three steps:

- Write a cover letter summarizing the unit.
- Choose papers to include from your work in this unit.
- Discuss your personal mathematical growth during the unit.

Cover Letter

Look back over *As the Cube Turns* and describe the central problem of the unit and the key mathematical ideas. Your description should give an overview of how the key ideas were developed and how they were used to solve the central problem.

In compiling your portfolio, you will select some activities you think were important in developing the unit's key ideas. Your cover letter should include an explanation of why you selected each item.

Selecting Papers from *As the Cube Turns*

Your portfolio for *As the Cube Turns* should contain these items:

- *How Did We Get Here?*
- *Spiders and Cubes*
- *Beginning Portfolio Selection* and *Continued Portfolio Selection*

  Include your own work on the activities from the unit that you selected in these activities.

continued ▶

- A Problem of the Week

  Select one of the first two POWs you completed during this unit (*"A Sticky Gum Problem" Revisited* or *A Wider Windshield Wiper, Please*).

- Your work on POW 9: *An Animated POW*

  Include your work from *An Animated Outline*.

Personal Growth

Your cover letter for *As the Cube Turns* describes how the mathematical ideas develop in the unit. In addition, write about your own personal development during this unit. You may also want to address this question:

> *How have you grown in your understanding of writing and interpreting programs?*

Include any other thoughts about your experiences that you wish to share with a reader of your portfolio.

SUPPLEMENTAL ACTIVITIES

The supplemental activities for *As the Cube Turns* continue the unit's areas of emphasis—programming, trigonometry, matrices, and transformational geometry—although other topics appear as well. Here are some examples:

- *Loopy Arithmetic* and *Let the Calculator Do It!* give additional work on writing programs.

- *Sum Tangents* and *Half a Sine* look at the development of other trigonometric formulas.

- *Bugs in Trees* shows a new application for matrices.

- *The General Isometry* is a challenging activity about combining isometries.

Loopy Arithmetic

The Basic For/End Loop

For/End loops can be used in programs to do a variety of repetitive tasks, including arithmetic. By combining this type of loop with a display of results on the screen, you can get your graphing calculator to show you the work it's doing.

For example, the screen for this plain-language program will show the calculator counting from 20 to 100:

> For A from 20 to 100
>
> > • Display A on the screen
>
> End the the A loop

Well, actually, you won't see most of the counting happen, because the numbers will whiz by on your screen. But when the program is finished running, it will look something like the screen shown here.

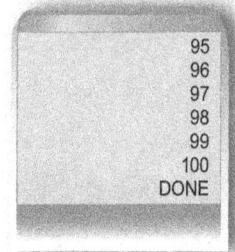

```
95
96
97
98
99
100
DONE
```

Beyond Counting

Your calculator can do more challenging tasks than simply counting. Here are two programs for you to write.

1. Write a program using a loop to compute factorials. Your program should ask the user for an input and then give the factorial of that number. The challenge of this program is writing it *without using the calculator's factorial instruction.*

2. You may be familiar with the **Fibonacci sequence** of numbers, which begins like this:

$$1, 1, 2, 3, 5, 8, 13, 21, 34, \ldots$$

In this sequence, each term is obtained by adding the two previous terms. For example, the term 34 comes from the sum $13 + 21$.

Write a program using a loop to compute and display the first 40 Fibonacci numbers.

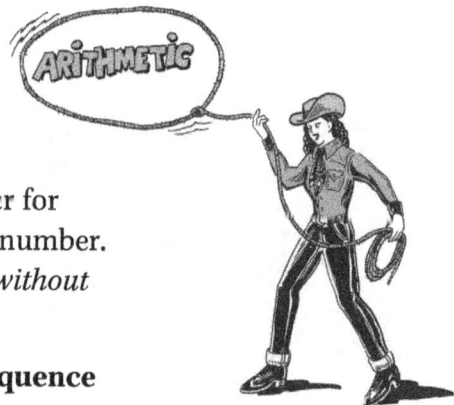

Sum Tangents

The sine and cosine functions are not the only functions for which there are angle-sum formulas. The tangent function has this angle-sum formula:

$$\tan (A + B) = \frac{\tan A + \tan B}{1 - \tan A \tan B}$$

Your task is to show how to prove this formula. One useful fact is the relationship $\tan \theta = \frac{\sin \theta}{\cos \theta}$.

1. a. Use the right-triangle definitions of the sine, cosine, and tangent functions to prove the relationship $\tan \theta = \frac{\sin \theta}{\cos \theta}$ for acute angles.

   b. Explain why this relationship holds true for all angles.

2. Use the relationship $\tan \theta = \frac{\sin \theta}{\cos \theta}$ and the angle-sum formulas for sine and cosine to prove the angle-sum formula for the tangent function.

3. Develop an angle-sum formula for the cotangent function similar to that for the tangent function. Here are two possible approaches.

   a. Apply the fact that $\cot \theta$ can be defined as $\frac{1}{\tan \theta}$ directly to the tangent-of-a-sum formula.

   b. Combine the fact that $\cot \theta$ is equal to $\frac{\cos \theta}{\sin \theta}$ with the angle-sum formulas for sine and cosine.

Moving to the Second Quadrant

In *The Sine of a Sum,* you proved this sine-of-a-sum formula:

$$\sin (A + B) = \sin A \cos B + \cos A \sin B$$

Because your proof depends on a diagram that requires A and B to be acute angles, it does not apply in all cases. However, the formula *does* hold true for all angles. In this activity, you will prove that it holds true when A is in the second quadrant (and B is still in the first quadrant).

The basic idea is to express A as 90° more than a first-quadrant angle. To use this idea, you need to develop formulas connecting the sine and cosine of a second-quadrant angle with the sine and cosine of the first-quadrant angle that is 90° less.

1. The first step is to find a formula for $\sin (\theta + 90°)$ in terms of either $\sin \theta$ or $\cos \theta$.

   a. Pick a value for θ and find $\sin (\theta + 90°)$. Then find $\sin \theta$ and $\cos \theta$, and write a general formula based on what you find.

   b. Verify your formula using other values for θ.

   c. Explain why your formula holds true for all values of θ, in these ways:

      • Using the graph from *Oh, Say What You Can See*

      • Using the Ferris wheel model

      • Using the general definition of the sine function in terms of coordinates

2. Develop and justify similar formulas for each expression.

   a. $\cos (\theta + 90°)$

   b. $\sin (\theta - 90°)$

   c. $\cos (\theta - 90°)$

continued ▸

3. Use your results from Questions 1 and 2 to find a formula for $\sin (A + B)$ for the case in which A is in the second quadrant but B is in the first quadrant.

To do this, write A as $x + 90°$ and show that $\sin (A + B)$ is equal to $\sin [(x + B) + 90°]$. Then apply the formula from Question 1 to express $\sin [(x + B) + 90°]$ as a trigonometric function of $x + B$. Next, use the appropriate function-of-a-sum formula to express this function of $x + B$ in terms of the sine and cosine of x and B. Finally, get back to A by using the fact that x is $A - 90°$ and applying appropriate formulas from Question 2.

Adding 180°

If you did the supplemental activity *Moving to the Second Quadrant,* you developed and explained formulas for the expressions $\sin(\theta + 90°)$ and $\cos(\theta + 90°)$ in terms of either $\sin\theta$ or $\cos\theta$. This activity is similar, except that you will be adding 180° to θ instead of adding 90°.

1. Find a formula for $\sin(\theta + 180°)$ in terms of either $\sin\theta$ or $\cos\theta$.

   a. Pick a value for θ and find $\sin(\theta + 180°)$. Then find $\sin\theta$ and $\cos\theta$, and write a general formula based on what you find.

   b. Verify your formula using other values for θ.

   c. Explain why your formula holds true for all values of θ, in at least one of these ways:

   - Using the graph from *Oh, Say What You Can See*
   - Using the Ferris wheel model
   - Using the general definition of the sine function in terms of coordinates

2. Develop and justify a similar formula for $\cos(\theta + 180°)$.

3. Explain how to use your answers to Questions 1 and 2 to get formulas for $\sin(\theta - 180°)$ and $\cos(\theta - 180°)$. That is, show how to develop formulas in which you subtract 180° instead of adding 180°.

Sums for All Quadrants

In the activity *The Sine of a Sum,* you showed that if A and B are first-quadrant angles, then $\sin(A + B)$ is equal to $\sin A \cos B + \cos A \sin B$. In the supplemental activity *Moving to the Second Quadrant*, you showed that this formula works when one of the angles is in the second quadrant and the other is in the first quadrant.

Your task now is to prove this formula for other cases.

1. Prove the formula for the case in which both A and B are second-quadrant angles. To do this, adapt the method used in *Moving to the Second Quadrant* by writing A as $x + 90°$ and B as $y + 90°$. You will also need a formula for $\sin(\theta + 180°)$ in terms of either $\sin \theta$ or $\cos \theta$.

2. Prove the formula for all other cases, considering the possible combinations of quadrants for A and B. Use the periodicity of the sine and cosine functions, and express angles in various quadrants in terms of appropriate first-quadrant angles.

   Begin by listing the cases you need to consider, and think about how to avoid duplication. For instance, if you do the case in which A is in the third quadrant and B is in the second quadrant, you do not have to also do the case in which these quadrants are reversed.

Deriving the Polar Complex

Some of the supplemental activities in the unit *The Diver Returns* introduced the polar form of a complex number and explained how this form is useful. To review briefly, if the rectangular form of the complex number z is $a + bi$, its polar form is $r(\cos \theta + i \sin \theta)$, where $r \geq 0$. This form follows from a graph of the number z in the complex plane.

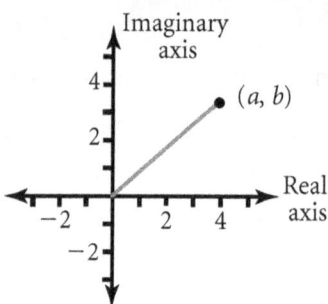

In the earlier problems, you were simply given formulas for products, quotients, and powers of complex numbers in polar form. Your task now is to derive those formulas by using the formulas you have found for the sine and cosine of the sum of two angles.

Suppose that $z_1 = r_1(\cos \theta_1 + i \sin \theta_1)$ and $z_2 = r_2(\cos \theta_2 + i \sin \theta_2)$ are two complex numbers written in polar form.

1. Use the formulas you have found for the sine and cosine of the sum of two angles to find a simple formula for the polar form of the product $z_1 \cdot z_2$.

2. Find a formula for the quotient $\frac{z_1}{z_2}$.

3. If n is a positive integer, find a formula for z_1^n.

Sine and Cosine Derivatives Again

In the unit *How Much? How Fast?* you developed formulas for the derivatives of the sine and cosine functions, using a geometric analysis. The formulas for the sine or cosine of the sum of two angles provide a different approach, which you will use in this activity. This approach is also based on the idea that for any function f, the derivative of f at $x = a$ is based on what happens to the derivative ratio $\frac{f(a + h) - f(a)}{h}$ as h gets closer to 0.

1. First consider the sine function.

   a. Apply the formula for the sine of a sum to the expression $\sin(a + h)$.

   b. Substitute your result into the derivative ratio for the sine function. Simplify the result so it looks roughly like this:
   $$f(a)\, g(h) + k(a)\, l(h)$$
   That is, simplify the result so it is the sum of two products, each of which has "an a part" and "an h part."

   c. Examine the two "h parts" of your expression to find out what happens to each of them as h gets closer to 0. (Remember to use radian measure.) For now, you can simply try smaller and smaller values for h and look for a trend, but explore Question 3 for a deeper analysis.

   d. Use your results to get an expression for the derivative of the sine function at $x = a$ that involves only a (and not h).

2. Use a similar sequence of steps for the cosine function. The two "h parts" you found in Question 1b should reappear here, so you can use your results from Question 1c again.

3. *Challenge:* In Question 1c, you probably used "very small" values for h to see note what happens to certain expressions as h gets closer to 0. Now look for a proof, perhaps using geometry and the definitions of the trigonometric functions, for why your results in Question 1c are correct.

Bugs in Trees

Matrices are a shorthand for representing certain numeric information, and matrix operations are a convenient way to describe certain arithmetic processes. In this unit, you've used matrices as a way of doing the arithmetic involved in geometric transformations, such as translations and rotations. In *Meadows or Malls?* (in Year 3), you used them to represent systems of linear equations.

This activity will illustrate another context in which matrices provide a useful way to describe patterns of arithmetic operations.

The Situation

Two trees, tree A and tree B, are side by side. Some bugs have infested the trees, and the bugs are moving back and forth. A researcher watches the bugs' activity, recording their movements once per hour, and reaches these conclusions:

• If a bug is in tree A at the time of a particular observation, there is a 70% chance that it will still be in tree A at the next observation and only a 30% chance that it will be in tree B at the next observation.

• If a bug is in tree B at the time of a particular observation, there is a 60% chance that it will be in tree A at the next observation and only a 40% chance that it will still be in tree B at the next observation.

In studying sequences of coin flips, you learned that each flip is independent of the previous flips. In other words, the probability of getting heads or tails on a given flip does not depend on what was flipped in the past.

In the situation here, the position of a bug at a given observation is analogous to the outcome of a given coin flip. But unlike the coin situation, a bug's position at a given observation is not independent of where it was before.

continued ▶

Specifically, the position of a bug at a given observation depends (in a probabilistic way) on where it was at the preceding observation. A situation in which each trial depends probabilistically on the preceding trial is called a *Markov chain*. The theory of Markov chains derives its name from the Russian mathematician A. A. Markov (1856–1922), who did pioneering work in this field.

1. Suppose that in a certain observation, there are 110 bugs in tree A and 90 bugs in tree B. If the movement of the bugs follows the researcher's analysis, how many bugs will be in each tree at the next observation?

2. Suppose the trees become infested with many thousands of bugs. At a given observation, 25% of the bugs are in tree A and 75% are in tree B. If these bugs continue to move according to the researcher's analysis, what fraction of them will be in each tree at the next observation?

The Matrices

The researcher's probabilities can be put into this matrix:

$$\begin{bmatrix} .7 & .3 \\ .6 & .4 \end{bmatrix}$$

This is called a *transition matrix,* because it describes how a bug's position might change from one observation to the next.

The results of a particular observation can be put into a row matrix. For example, in Question 1, the information obtained can be represented by the row vector $[110 \quad 90]$.

3. Show why the arithmetic you did in Question 1 can be represented in matrix form by this product:

$$[110 \quad 90] \cdot \begin{bmatrix} .7 & .3 \\ .6 & .4 \end{bmatrix}$$

4. Show how to represent the arithmetic of Question 2 as the product of a row vector and a matrix.

continued ◈

The Hours Go By

For the rest of this activity, continue to assume that the researcher's analysis holds true. Start with the situation from Question 2, and suppose that you make an observation every hour. Treat the initial situation as hour 0 and your result from Question 2 as hour 1.

5. Find the percentage of bugs in each tree at hour 2. Remember, a bug's position at hour 2 depends on where it was at hour 1, but not on where it was at hour 0.

6. Develop a matrix expression for the distribution of the bugs at hour n.

7. What will happen in the long run? Will there ever come a time when the bugs are all in tree A? Explain your answers.

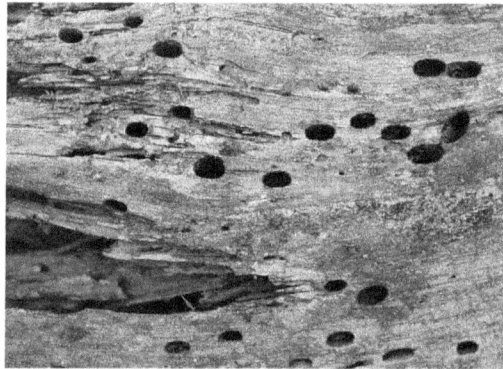

Adapted with permission from the *Mathematics Teacher*, © May 1998, by the National Council of Teachers of Mathematics.

Half a Sine

You have found formulas that allow you to find the sine and cosine of the sum and the difference of two angles in terms of the sines and cosines of the two angles themselves. You also found formulas for double angles in *Doubles and Differences.*

Your task here is somewhat the opposite. You are to find formulas for the sine and cosine of half of a given angle in terms of the sine and cosine of the angle itself. In other words, you want to find formulas that look like this:

$$\sin\left(\frac{A}{2}\right) = \ldots$$

$$\cos\left(\frac{A}{2}\right) = \ldots$$

In each case, the right side of the equation can involve any trigonometric functions using the angle A itself.

Start by thinking of A as being twice $\left(\frac{A}{2}\right)$, and use the double-angle formula to write $\cos A$ in terms of $\sin\left(\frac{A}{2}\right)$ and $\cos\left(\frac{A}{2}\right)$. Then use the Pythagorean identity, $\sin^2\theta + \cos^2\theta = 1$, to get $\cos A$ in terms of only $\sin\left(\frac{A}{2}\right)$ or only $\cos\left(\frac{A}{2}\right)$. Work from that to get $\sin\left(\frac{A}{2}\right)$ and $\cos\left(\frac{A}{2}\right)$ in terms of $\cos A$.

The General Isometry

Mathematicians often use the word *transformation* to indicate a function in which the domain and range are sets of points. If we use the letter f to represent a specific transformation, then $f(X)$ represents the point to which X is moved.

An *isometry* is a special type of geometric transformation—one in which the distance between two points remains unchanged. In other words, if f is an isometry and A and B are any two points, then the distance from $f(A)$ to $f(B)$ must equal the distance from A to B.

You've also seen that there are three fundamental types of isometries of the plane: translations, rotations, and reflections. In this activity, suppose that ABC is a triangle and that f is an isometry of the plane. Your task is to show, as described in Questions 1 to 3, that f can be created by combining the three basic types of isometries.

1. Suppose RST is a triangle that is congruent to triangle ABC, with $AB = RS$, $BC = ST$, and $AC = RT$. (*Reminder:* The notation XY means the distance from X to Y.)

    Show that one of these two statements must be true:

    *   There is a translation g, a rotation h, and a reflection k for which $R = k(h(g(A)))$, $S = k(h(g(B)))$, and $T = k(h(g(C)))$.
    *   There is a translation g and a rotation h for which $R = h(g(A))$, $S = h(g(B))$, and $T = h(g(C))$.

    First show that there is a translation g for which $g(A) = R$. Then show that there is a rotation h around R for which $h(g(B)) = S$. Finally, decide if a reflection across the line RS is needed.

2. Suppose RST is a triangle, and suppose that points X and Y satisfy these three conditions:

    * $RX = RY$
    * $SX = SY$
    * $TX = TY$

    Show that X and Y are actually the same point.

continued ▶

3. Use your results from Questions 1 and 2 to prove that one of these two statements must be true:

- There is a translation g, a rotation h, and a reflection k such that $f(X) = k(h(g(X)))$ for every point X.

- There is a translation g and a rotation h such that $f(X) = h(g(X))$ for every point X.

In other words, show that the isometry f is either a combination of a translation, a rotation, and a reflection *or* a combination of only a translation and a rotation.

Perspective on Geometry

One of the key tasks in this unit is deciding how to represent a three-dimensional object—a cube—on a two-dimensional screen.

Artists use the general term *perspective* to describe the various methods they use to create this type of representation. Your task in this activity is to conduct some research on the history of perspective in art.

In your report, describe different schemes of perspective and explain the geometric principles behind them. You may want to include examples of famous works of art showing different methods or create your own drawings illustrating how the same object might be drawn using different methods.

"Venice: A Regatta on the Grand Canal" by the Italian painter Canaletto (1697–1768)

Let the Calculator Do It!

In *Find Those Corners!* you projected the vertices of a cube onto the plane $z = 5$. You probably found that computing the projections for each vertex was a lot of work.

In this activity, you will write a program for your calculator to do that work.

Your program should ask the user for the vertex of the cube or for any point the user wants projected. It should also ask the user for the viewpoint. The program should then tell the user the projection of the given point, using the given viewpoint, onto the plane $z = 5$.

PHOTOGRAPHIC CREDITS

Front Cover Photography

(upper row) Stephen Loewinsohn; (lower row) iStockphoto

As the Cube Turns

1 (upper row) Stephen Loewinsohn; (lower row) iStockphoto; **3** Peter Jonnard, Hillary Turner, Richard Wheeler; **4** Stephen Loewinsohn; **8** Stephen Loewinsohn; **10** Hillary Turner; **11** Anthony Pepperdine; **12** iStockphoto; **14** Hillary Turner; **18** iStockphoto; **19** iStockphoto; **20** Shutterstock; **21** Terry Nowak, Lynne Alper; **26** iStockphoto; **27** Superstock, Inc.; **28** Peter Jonnard, Hillary Turner, Richard Wheeler; **29** Shutterstock; **35** iStockphoto; **36** iStockphoto; **39** iStockphoto; **43** iStockphoto; **45** Dave Robathan; **47** iStockphoto; **49** iStockphoto; **55** iStockphoto; **59** iStockphoto; **66** Photodisc; **68** Chicha Lynch, Hillary Turner, Richard Wheeler; **71** iStockphoto; **72** iStockphoto; **74** Stephen Loewinsohn; **90** Shutterstock; **93** iStockphoto; **94** Wikimedia Commons

www.ingramcontent.com/pod-product-compliance
Lightning Source LLC
Chambersburg PA
CBHW051227200326
41519CB00025B/7271